Animals of Africa

Lion

I0875896

Katie Gillespie

AV2
www.av2books.com

AV2

Step 1
Go to **www.av2books.com**

Step 2
Enter this unique code
YUOXQ1GYZ

Step 3
Explore your interactive eBook!

AV2 is optimized for use on any device

Your interactive eBook comes with...

Contents
Browse a live contents page to easily navigate through resources

Audio
Listen to sections of the book read aloud

Videos
Watch informative video clips

Weblinks
Gain additional information for research

Try This!
Complete activities and hands-on experiments

Key Words
Study vocabulary, and complete a matching word activity

Quizzes
Test your knowledge

Slideshows
View images and captions

This title is part of our AV2 digital subscription

1-Year Grades K–5 Subscription
ISBN 978-1-7911-3320-7

Access hundreds of AV2 titles with our digital subscription.
Sign up for a FREE trial at **www.av2books.com/trial**

Lion

CONTENTS

Meet the Lion

Lions are fierce and strong cats. They live on **savannas** or grassy plains. Lions prefer **habitats** with large amounts of open space because these areas are good for hunting.

In the past, lions had habitats in most of Africa. They could also be found as far north as Greece and as far east as India. Today, lions mainly live in the south Sahara Desert, and in southern and eastern Africa. A small population of about 650 Asiatic lions also lives in India's Gir Forest.

A Big Cat

Lions are one of the biggest cats in the world. A lion's tail alone can measure almost 40 inches (102 centimeters) in length. Adult male lions grow 6.5 to 9 feet (2 to 2.75 meters) long. Female lions, or lionesses, are smaller. They can be 4.5 to 5.5 feet (1.4 to 1.7 m) in length.

Lions typically weigh between 265 and 420 pounds (120 and 190 kilograms). This means that a lion can become even larger than an NFL player!

Size Difference

TIGER
10.8 feet
(3.3 m)

LION
9 feet
(2.7 m)

CHEETAH
7 feet
(2.1 m)

Lion Pride

Lions are very **social** animals. They typically live in groups, or prides, of 15 or more. Prides are formed around family connections. In each pride, a lion's rank is based on its age. When eating, adult male lions eat first, followed by adult female lions. The youngest lions eat last.

Small groups of male lions lead prides. Eventually, younger males will challenge them for control of the pride.

Male lions may fight to the death for control of a pride.

BIG FACT

A **lion pride** can have up to **40 members**.

Growing Up

Female lions give birth away from their prides. They introduce their babies, or cubs, to the pride when the cubs are about six weeks of age. Lion cubs are born with splotchy marks or stripes on their coats.

Male lions live with their prides until they are about two years old. Lionesses usually stay with the same pride for their whole lives. Lions can live about 15 years in nature.

Lionesses usually give birth to two to four cubs at a time.

A Closer Look

Lions have **adapted** to their habitats. They have unique features to help them survive in their environment.

Mane

A male lion's shaggy mane signals his strength and power to other lions.

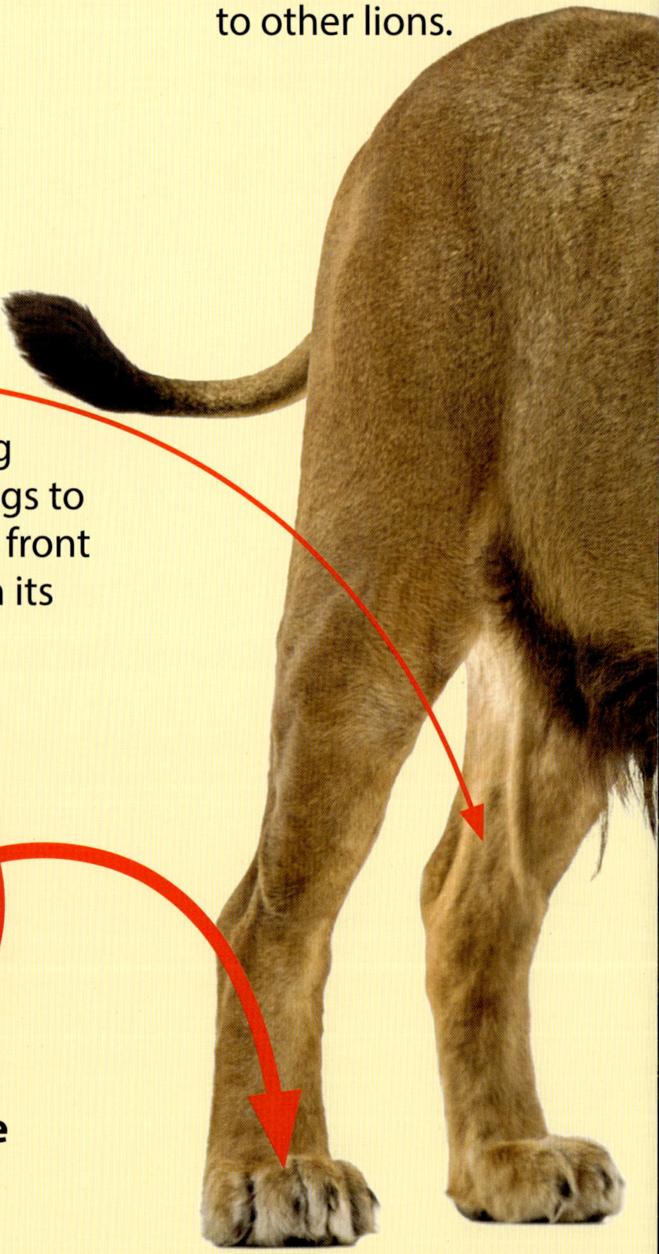

Legs

A lion uses the strong muscles in its back legs to leap toward **prey**. Its front legs are used to push its prey to the ground.

Paws

A lion's paws have **retractable** claws and soft pads that help the animal move quietly while hunting.

Eyes

A lion's large eyes help it to see well at night.

Mouth

A lion's powerful jaws and long, pointed teeth are used for killing other animals.

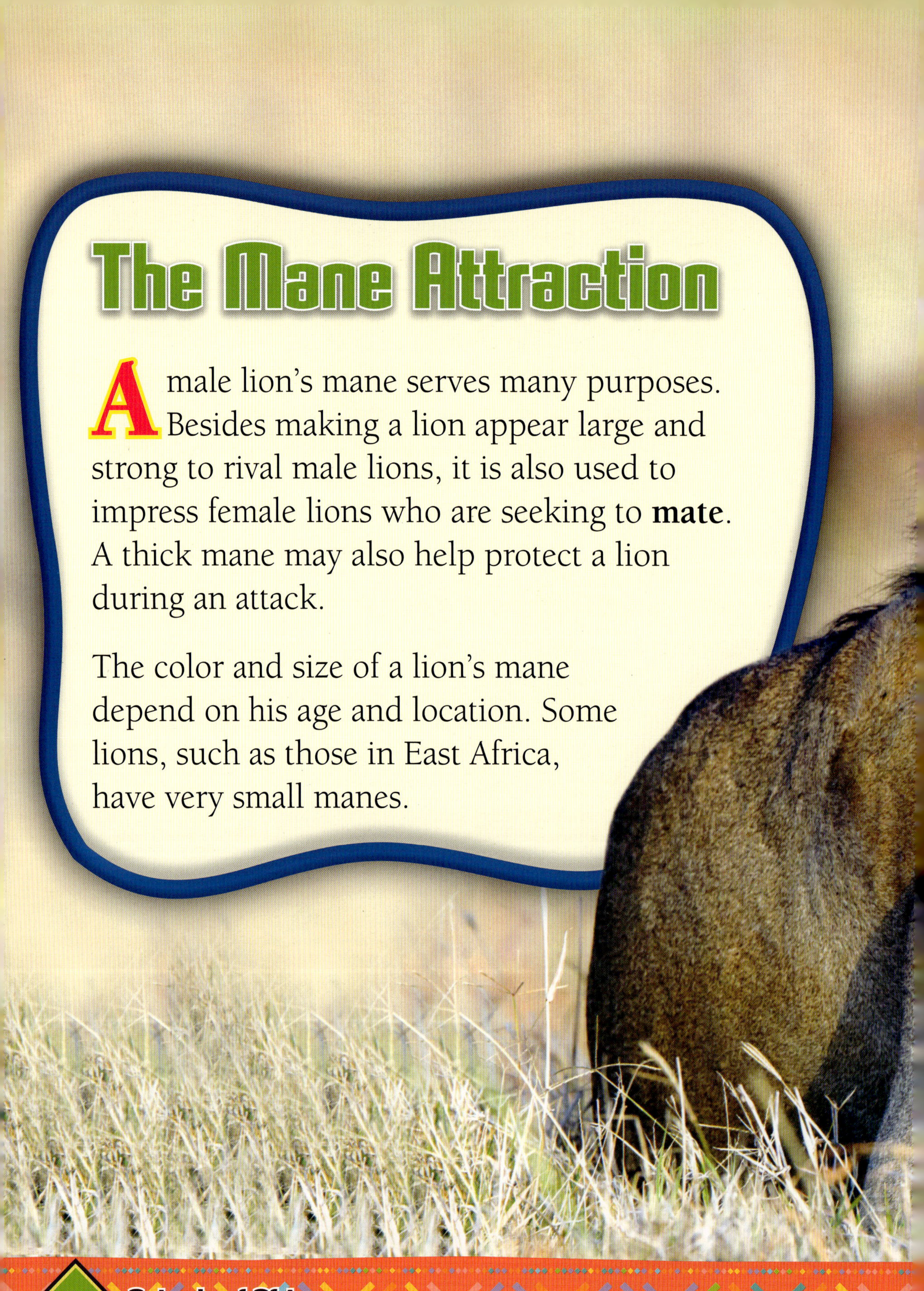

The Mane Attraction

A male lion's mane serves many purposes. Besides making a lion appear large and strong to rival male lions, it is also used to impress female lions who are seeking to **mate**. A thick mane may also help protect a lion during an attack.

The color and size of a lion's mane depend on his age and location. Some lions, such as those in East Africa, have very small manes.

BIG FACT

Male lion cubs start growing a mane after their **1st birthday**.

Hear Me Roar

Lions roar to **communicate** with other lions. A lion's roar is very loud and can travel great distances. In fact, a roar can be heard from 5 miles (8 kilometers) away.

The sound of a roar is meant to keep rival lions away from a pride's **territory**. Lions also roar in order to call out to members of their own pride. Roaring helps the family stay in touch. Lions do not roar when they are hunting because they do not want to spook their prey.

Getting Loud
Jet Takeoff
150 decibels
Chain Saw
120 decibels
Lion's Roar
114 decibels
Vacuum Cleaner
70 decibels
Human Voice
60 decibels

Night Hunters

Lions are nocturnal. This means that they are most active at night. Female lions do most of the hunting for the pride. While male lions guard the pride's territory and cubs, the lionesses hunt. Working together, they separate their prey from its herd. Next, they chase it toward the other hunters. Since they are lighter than male lions, lionesses can run very quickly, up to 50 miles (80 km) per hour.

Lions often prey on large animals such as buffalo, zebras, and wildebeests.

Lions in Danger

Lions hunt in large areas. As humans continue to expand their **development** of the landscape for farming, more and more lion habitat is taken away. The food that lions find in nature, such as zebras and gazelles, is also disappearing. Sometimes, farmers kill or poison lions to prevent them from eating their **livestock**. All of these activities combined have shrunk the population of lions living in Africa today.

Tourism, and the money it brings, is helping to save lions and their habitat.

Population Decline*
100,000
in 1960
20,000
in 2020
*Estimate

Lion Quiz

1 How long can a lion's tail measure?

2 What is a group of lions called?

3 How many cubs does a lioness usually give birth to at a time?

4 What features on a lion's paws help the animal hunt quietly?

5 When do male lions start growing their manes?

6 How far away can a lion's roar be heard?

7 When are lions most active?

8 About how many lions are found in nature today?

ANSWERS **1.** Almost 40 inches (102 cm) **2.** A pride **3.** Two to four **4.** Retractable claws and soft pads **5.** After their first birthday **6.** 5 miles (8 km) **7.** At night **8.** About 20,000

Key Words

adapted: changed to suit the environment

communicate: to exchange information, news, or ideas

development: the placement of structures on land for human use

habitats: the places where animals live, grow, and raise their young

livestock: animals kept on a farm for use or profit

mate: to come together for breeding

prey: an animal hunted or captured by another animal for food

retractable: able to be pulled back in

savannas: flat areas of land with few trees

social: preferring the company of others

territory: a region that is defended by an animal

Index

Get the best of both worlds.

AV2 bridges the gap between print and digital.

The expandable resources toolbar enables quick access to content including **videos**, **audio**, **activities**, **weblinks**, **slideshows**, **quizzes**, and **key words**.

Animated videos make static images come alive.

Resource icons on each page help readers to further **explore key concepts**.

Published by AV2
276 5th Avenue
Suite 704 #917
New York, NY 10001
Website: www.av2books.com

Copyright ©2022 AV2
All rights reserved. No part of this publication may be reproduced, stored in a retrieval system, or transmitted in any form or by any means, electronic, mechanical, photocopying, recording, or otherwise, without the prior written permission of the publisher.

Library of Congress Cataloging-in-Publication Data
Names: Gillespie, Katie, author.
Title: Lion / Katie Gillespie.
Description: New York, NY : AV2, [2022] | Series: Animals of Africa | Includes index. | Audience: Ages 8-11 | Audience: Grades 2-3 | Summary: "Learn about some of the world's most fascinating animals in Animals of Africa"-- Provided by publisher.
Identifiers: LCCN 2020053940 (print) | LCCN 2020053941 (ebook) | ISBN 9781791135270 (library binding) | ISBN 9781791135287 (paperback) | ISBN 9781791135294
Subjects: LCSH: Lion--Juvenile literature.
Classification: LCC QL737.C23 G51416 2022 (print) | LCC QL737.C23 (ebook) | DDC 599.757--dc23
LC record available at https://lccn.loc.gov/2020053940
LC ebook record available at https://lccn.loc.gov/2020053941

Printed in Guangzhou, China
1 2 3 4 5 6 7 8 9 0 25 24 23 22 21

012021
101120

Project Coordinator: Heather Kissock
Designer: Terry Paulhus

Photo Credits
Every reasonable effort has been made to trace ownership and to obtain permission to reprint copyright material. The publisher would be pleased to have any errors or omissions brought to its attention so that they may be corrected in subsequent printings.
AV2 acknowledges Getty Images, Minden Pictures, and iStock as its primary image suppliers for this title.

View new titles and product videos at www.av2books.com